NONFICTION

URANUS

URANUS

SEYMOUR SIMON

WILLIAM MORROW AND COMPANY, INC.
New York

PICTURE CREDITS:
All photographs courtesy of
the Jet Propulsion Laboratory
(California Institute of Technology)/NASA.
Drawing on page 8 by Todd Radom;
artwork on pages 9 and 15 by Frank Schwarz.

Printed in Italy.

1 2 3 4 5 6 7 8 9 10

Library of Congress Cataloging-in-Publication Data
Simon, Seymour.
Uranus.
Summary: Introduces, through text and photographs, the
characteristics of the seventh planet in the solar
system.
1. Uranus (Planet)—Juvenile literature. [1. Uranus
(Planet) 2. Planets] I. Title.
QB681.S56 1987   523.4'7   86-31223
ISBN 0-688-06582-1
ISBN 0-688-06583-X (lib. bdg.)

Remembering my beloved sister
Roslyn Simon Orlofsky

On the evening of March 13, 1781, William Herschel stood in his garden in Bath, England. Herschel, a musician and amateur sky-watcher, was looking through his small telescope when he noticed a tiny ball of light that had wandered from where he had last seen it among the stars. It was a new planet, circling the sun far beyond what was thought to be the edge of the Solar System. Years after its discovery, astronomers named the planet Uranus, after the Greek god of heaven and the ruler of the world.

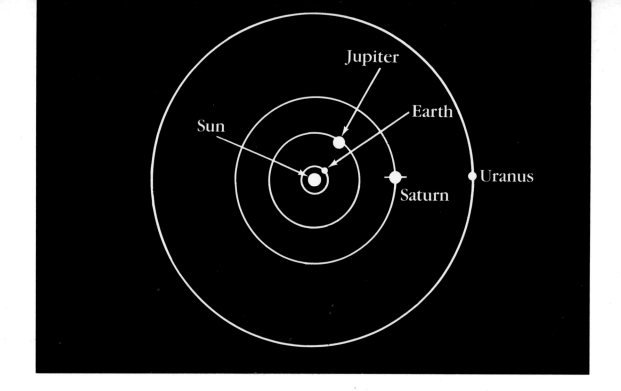

Uranus [YUR-uh-nus] is the seventh planet in the Solar System. It is about one and three-quarter billion miles from the sun, nineteen times farther away than our planet Earth. Uranus is so far away that it takes eighty-four years to circle the sun once.

Like Jupiter, Saturn, and the eighth planet, Neptune, Uranus is a giant planet made up mostly of gases. But Jupiter and Saturn are far larger than Uranus. About thirty-one thousand miles across, Uranus is halfway in size between Jupiter and Earth. If Uranus were hollow, about fifty planet Earths could fit inside.

Jupiter

Uranus

Earth

Neptune

Saturn

In January 1986, the *Voyager 2* spacecraft swept past Uranus, eight and one-half years after it had been launched from Earth and more than four years after it had viewed Saturn. This photograph of Uranus was taken when *Voyager 2* was still more than five million miles from the planet.

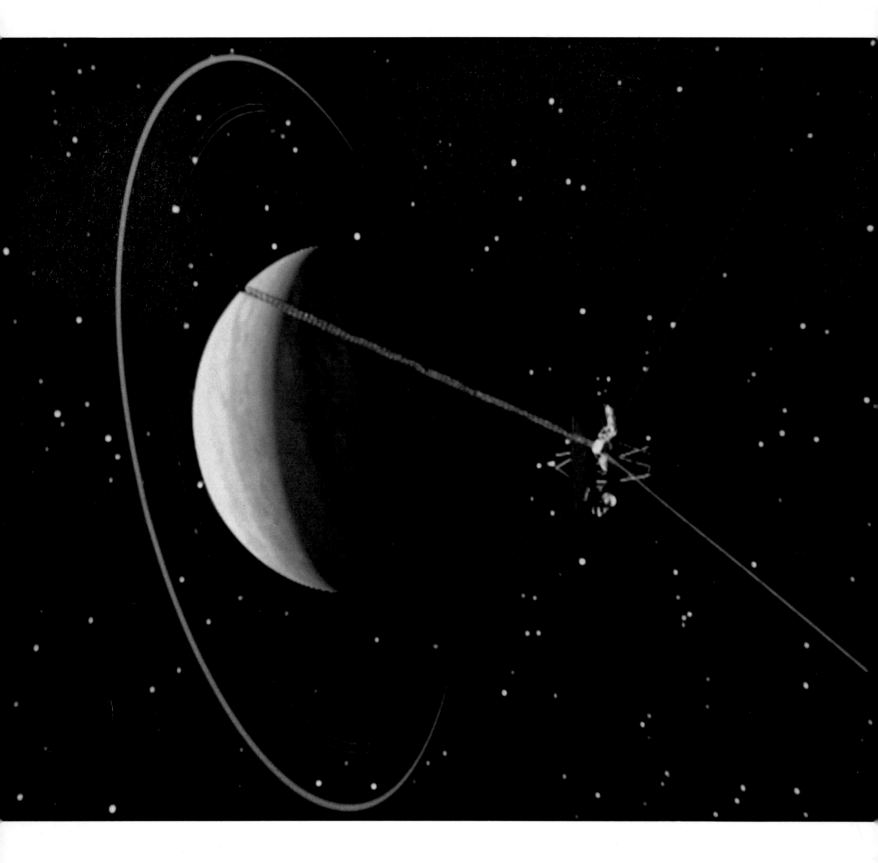

In this computerized drawing, *Voyager 2* is shown at its closest approach to Uranus on January 24, 1986. The tops of pale blue-green clouds are only fifty-one thousand miles away.

Compared to the other planets in the Solar System, Uranus "lies on its side" in space. Scientists think that Uranus was once knocked over on its side by a collision with another object in space. At the moment, Uranus's south pole is pointing at the sun. A visitor there would always see the sun shining directly overhead. In fact, the south pole is in the midst of forty-two years of constant sunlight, while the north pole is having a forty-two-year night. Gradually, after these years of darkness, the north pole will point at the sun and it will have forty-two years of daylight, while the south pole will be dark.

Uranus is surrounded by a strange magnetic field. Most planets have magnetic fields lined up with their north and south poles. But Uranus's magnetic field is closer to its equator than it is to the poles. This makes the magnetic field wobble back and forth as the planet spins.

The magnetic field's shape is equally strange. It has a long, gently curving tail on the side away from the sun. The tail keeps twisting and turning as the planet moves.

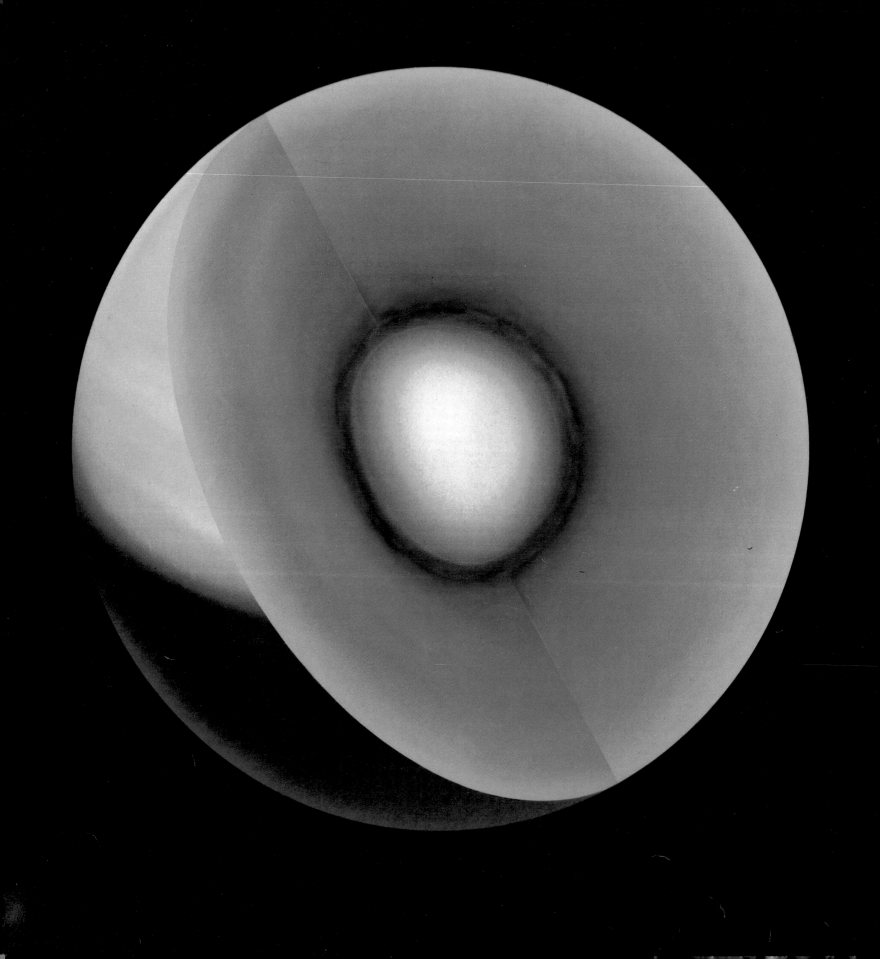

Imagine yourself flying in a spaceship high over the planet. Around you is a thin layer of haze and smog. Below is an atmosphere of hydrogen and helium gas. Thick blue-green clouds blow across the surface at two hundred miles per hour. The clouds are blue-green because they contain methane gas, which looks green in the sunlight. Temperatures at the cloud tops are over 350 degrees (F) below zero—hundreds of degrees colder than the coldest spot on Earth.

As your spaceship sinks lower, the atmosphere gets hotter. Boiling drops of superheated water and steam are all around you as the temperature climbs thousands of degrees. Some scientists now think that the atmosphere of Uranus may become heavier and more liquid for twelve thousand miles—but no one knows for sure. Perhaps Uranus has no "surface" at all, just a hot, watery atmosphere down to its molten, rocky core about the size of planet Earth. This drawing shows the hazy atmosphere around the core.

Before the *Voyager 2* flyby, astronomers thought Uranus had nine rings. However, two new rings were discovered, along with parts of other rings.

Saturn's rings are filled with fine dust particles. But the Uranian rings are mostly dust free. Instead, they are made of spinning chunks of an unknown black material. Ranging in size from three to three thousand feet across, these chunks look like lumps of coal on a merry-go-round.

This photograph shows nine of the rings from two and one-half million miles away. The bright ring at the top is called epsilon, which is a Greek letter.

This computer-colored photograph shows one part of Uranus's epsilon ring. In reality, the ring is very dark. The color makes it possible to see some of the details. The epsilon ring contains narrow bands less than one hundred feet thick and lying about one mile apart. This slice of the ring is about twenty miles wide. Other parts of the ring are only fourteen miles wide. Scientists think that the varying width is caused by the gravitational pull of small moons moving nearby.

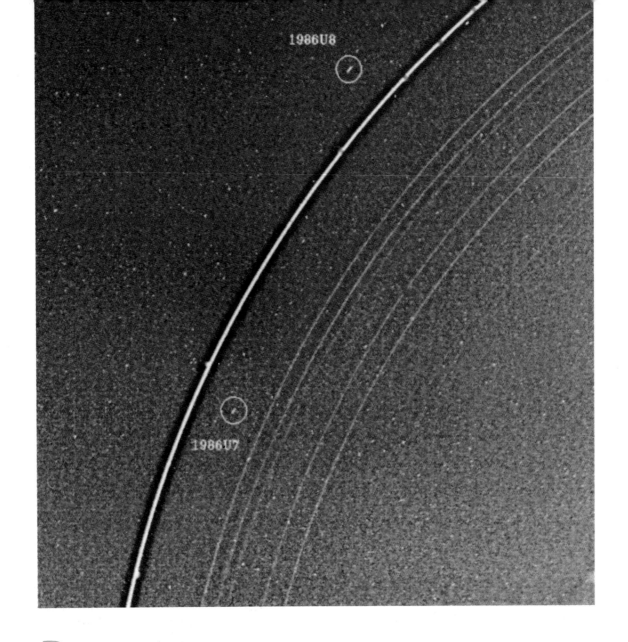

1986U8

1986U7

Before *Voyager 2*, Uranus was known to have five moons. *Voyager* discovered ten new moons, most about thirty miles across. In this photo you can see two of the new moons traveling on either side of the epsilon ring.

This photo shows three more new moons farther away from the epsilon ring. The ten new moons are now called 1985U1 and 1986U1 through 1986U9. These are temporary names using the year and the order in which they were discovered. Within a few years, the moons will be given their permanent names by the International Astronomical Union, a worldwide association of astronomers first organized in 1919.

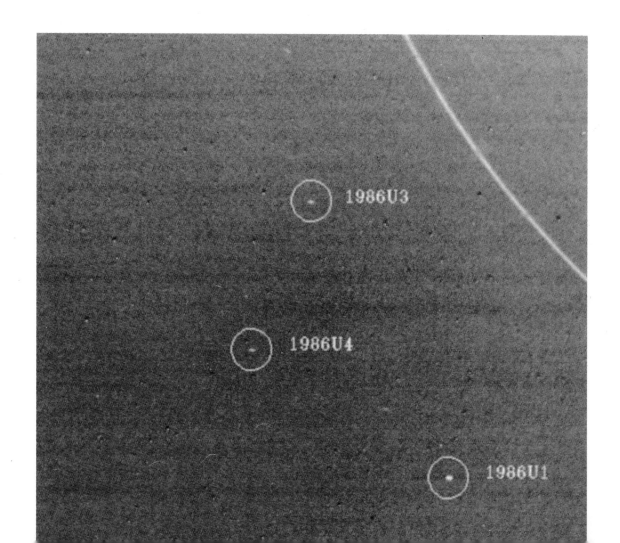

The five earlier known moons had been named after characters in English literature. Oberon and Titania, the two largest moons of Uranus, were discovered by William Herschel in 1787. Herschel's son named them after the king and queen of fairyland in Shakespeare's play *A Midsummer Night's Dream.*

Oberon [O-buh-ron], the outermost moon, is about 950 miles across and circles Uranus at a distance of 350 thousand miles. The *Voyager 2* photograph shows a tall mountain and many craters on its icy surface. The larger craters were once flooded by lava that has now become dark rock.

Titania [Ty-TAY-ne-ah], the next closest moon, is a bit larger than Oberon and circles Uranus at a distance of 272 thousand miles. Titania is a grayish moon with many bright ray craters—round, saucer-like pits—and long cracks. The surface of Titania is covered by very rough rocks probably broken apart by a bombardment of tiny meteorites. This view of Titania shows part of a thousand-mile-long trench near the day-night boundary at the left. At the top, you can see a large basin that was caused by a collision with a giant meteorite sometime in the past.

Gray Umbriel [UM-bree-ul] is the darkest of the five large moons. Its surface is covered by overlapping craters and impact basins. A puzzling bright ring about ninety miles across can be seen near the top of the photo. Called the "fluorescent Cheerio," it appears to be the floor of a crater. Umbriel is about seven hundred miles across and circles Uranus at a distance of 160 thousand miles.

Ariel [AIR-ee-ul] is the brightest of all of the moons. Broad, curving valleys and huge canyons cut across its surface. The bright spots are the rims of small craters. At one time in the distant past, water or ice might have flowed over the moon, filling the valleys with rocky material. Ariel is about the same size as Umbriel and circles Uranus at a distance of 115 thousand miles. These two moons were named after characters in a poem by the English writer Alexander Pope.

Imagine yourself preparing to land on the surface of Miranda [mi-RAN-duh], innermost of the five larger moons. The blue-green cloud tops of Uranus are only sixty-five thousand miles away. On the right is one of the huge canyons that cover the surface of Miranda. This view was made by combining several photographs and a drawing of the rings.

Miranda is the strangest of all five moons. Its surface has deep grooves and ridges and is covered by rope-like markings that look like the brushwork of some gigantic artist. The *Voyager 2* scientists named the oval racetrack pattern (bottom right) for the "circus maximus" of ancient Rome, where chariot races were held. The scientists think Miranda has the oddest mixture of surfaces ever found in the Solar System.

Close-up views of Miranda show the torn crust and the grooves that make up the "great racetracks" on the surface. The "racetracks" seem to be rows of long valleys or cracks ten times deeper than the Grand Canyon—all this on a moon only three hundred miles across.

This farewell view of Uranus was taken as *Voyager 2* left the planet behind and continued out into space toward a meeting with mysterious Neptune in 1989. In just the few short days of *Voyager*'s flyby, scientists had learned much about Uranus, its moons, and its rings. But they will be puzzling over the planet's many remaining mysteries for a long time to come.